Dr Henry Hicks
(1837-99)

The Life and Times of Dr Henry Hicks of St David's, and the Bubble that Refused to Burst

T0300820

Dyfed Elis-Gruffydd

First impression: 2023

© Dyfed Elis-Gruffydd & Y Lolfa Cyf., 2023

This book is subject to copyright and may not be reproduced by any means except for review purposes without the prior consent of the publishers.

ISBN: 978-1-80099-331-0

Published and printed in Wales on paper
from well-maintained forests by
Y Lolfa Cyf., Talybont, Ceredigion, SY24 5HE
email ylolfa@ylolfa.com
website www.ylolfa.com
tel 01970 832 304
fax 01970 832 782

Acknowledgements

I wish to express my thanks to Y Lolfa Press, and in particular to Garmon Gruffudd and Richard Huw Pritchard for the design and printing of this publication.
Thanks also go to my wife, Siân, for her support and advice and to the Sedgwick Museum of Earth Sciences, University of Cambridge for providing a photographic portrait of Dr Henry Hicks.

Foreword

I FIRST ENCOUNTERED Henry Hicks, or at least the name, during my days as a research student at University College, London. Although my particular research topic was the glacial history of the Brecon Beacons, Fforest Fawr and Bannau Sir Gâr, I had a passing interest in the literature relating to glacial deposits to be found in the London area in general, and particularly those in the vicinity of University College.

Amongst the papers read was one in the *Quarterly Journal of the Geological Society* of 1892 entitled 'On the Discovery of Mammoth and other Remains in Endsleigh Street and on Sections exposed in Endsleigh Gardens, Gordon Street, Gordon Square, and Tavistock Square, London' – to quote the convoluted title of the paper. I can recall how impressed I was by the way that the author had amassed and interpreted his evidence, and correctly concluded that the fossil bones and associated plant remains found in the gravels underlying the streets at the back of University College were proof of the 'Migration of the Mammoth and other northern animals southwards in advance of an ice-sheet, or of glaciers radiating from northern and north-western centres.'

The author in question was 'Henry Hicks, Sec. Geol. Soc.', that is, Secretary of the Geological Society of London established in 1807, during the so-called 'Heroic Age of Geology'. Further background reading relating to my own research in Wales revealed that Hicks was also the author of two papers on the evidence of ice action in

Pembrokeshire, published in 1892 and 1894, respectively, and a collection of very important articles on the bone caves of the Vale of Clwyd, which appeared between 1884 and 1888.

But it was not until I settled in Pembrokeshire in 1973 that I began to read Hicks' papers on the geology of the St. David's peninsula. And, much to my shame, it was not until I read a brief biographical statement in O.E. Roberts' book, *Rhai o Wyddonwyr Cymru* [Some of Wales' Scientists], published in 1980, that I learned that Henry Hicks was not only a Welshman, who had been born and brought up in St David's, but that this medical practitioner by profession was also a very accomplished 'amateur' geologist. And, according to Roberts, his interest in geology had been inspired 'by an officer of the Geological Survey [Salter] who was studying the fossils of his home area'.

Dyfed Elis-Gruffydd

Dr Henry Hicks (1837-99)

Henry Hicks (1837-99) a thorn in the side of the geological establishment and the bubble that refused to burst

B Y THE MID-NINETEENTH century, Henry Hicks was a prominent figure in the St David's area. Furthermore, he was without doubt a man of many talents and interests. On August 9th, 1870, the first of a series of *Literary and Musical Entertainments*, to quote the correspondent of the *Dewsland & Kemes Guardian*, the weekly newspaper which was printed at Martha Whiteside Williams' printers in Solfach (SM 801243), was held. Amongst the 15 singers and elocutionists who received a very warm response from the audience, that had assembled in the St David's Town Hall, was the local doctor and chemist Dr Henry Hicks, who recited with gusto a recitation titled 'A law suit'! On 1st September at the second of such evenings Hicks took to the stage, not once but three times and not to recite on this occasion but to sing two solos and a quartet. The evening came to a close with the 33 year old doctor enthusiastically singing 'Tight little island'.

The following week his service was called upon again. This time at the Grand Amateur Concert held in the Town Hall to raise money for the Cathedral Renovation Fund, a task undertaken under the supervision of Sir George Gilbert Scott, the famous architect whose work is evident in the Cathedral's present structure.

St David's Cathedral

Hicks not only supported the Cathedral's fund-raising but on 7th November he was called upon to treat John Williams, a worker who had badly injured himself as a result of falling to the Cathedral's floor from its ruinous walls. This particular accident occured at an inconvenient time for, in his role as treasurer, Hicks was in the process of collecting subscriptions for the St David's Reading Room, that was to be opened the following day. Hicks was also Honorary Secretary of the St David's Lifeboat, a rowing-boat named *Augusta* of which his uncle, David Hicks, was helmsman.

When he wasn't being called upon to treat patients, entertain the inhabitants of St David's, act as treasurer and secretary of community ventures, Hicks attended every possible public meeting where he would express his opinion on a variety of subjects that were of considerable importance to him, such as the education of the young of the area and the possibility of attracting the railway to St David's, a cause of which he was strongly in favour, although his hopes were not fulfilled.

Clearly, Hicks was interested in all aspects of the social and cultural life of St David's and he expressed great concern for the health of its inhabitants. However, his abiding interest was geology, especially the task of deciphering the truly ancient rocks of the St David's peninsula.

*   *   *

Hicks' family on his father's side hailed from Tre-maen-hir (SM 828262), a fine old stone-built house in the vicinity of Felin Ganol, near Solfach. The property was purchased by Abel Hicks, Henry Hicks' great-grandfather in 1752, and subsequently it was the home of his grandfather, Henry, and father, Thomas.

Hicks, however, was born, not at Tre-maen-hir, a house he invariably named and located on his maps of the geology of the St David's peninsula, but in St David's itself on 26th May 1837. Sadly, his father, Thomas Hicks, Surgeon (the local doctor) was granted little time in the company of his two young sons, Henry, 6, and John, 5, for soon after the birth of his third child, his daughter Mary Anne, Thomas Hicks died of the "decline" – probably tuberculosis – on 12th October 1843 and was buried amongst other members of the family in the graveyard of Eglwys Dewi Sant, Tre-groes, Solfach (SM 799254), situated some two miles west of Tre-maen-hir.

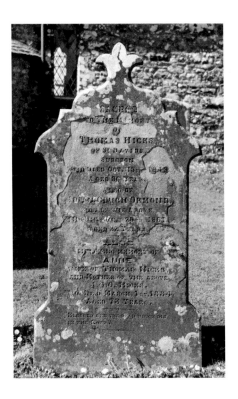

Anne Hicks, Thomas' widow, was now solely responsible for raising her young family. Although no records remain, we know that Henry was educated at the Cathedral Chapter School in St David's and despite losing his father at the tender age of 6, he decided to follow in Thomas Hicks' footsteps, for at the age of 22 he entered Guy's Hospital Medical School, London, in October 1859. Three years later, in March 1862, he gained his Licentiate of the Society of Apothecaries, and in April 1862, his MRCS (Membership of the Royal College of

Surgeons). Although he did not seek the higher qualification of FRCS (Fellow of the Royal College of Surgeons), he was highly regarded by such eminent practitioners as Dr Thomas Stevenson, FRCP (Fellow of the Royal College of Physicians), Examiner of Forensic Medicine at the University of London and Lecturer in Chemistry at Guy's Hospital, who, in October 1877, prepared one of three testimonials required by Hicks when he took his MD at St Andrews University in 1878. Stevenson recalls that Hicks '. . . was a highly successful and distinguished student and obtained very high positions at the Hospital prize examinations. Mr Hicks has been, and is, a successful practitioner, highly esteemed by all who know him. He is regarded by his professional brethren as a very able man in his profession'. And Thomas Stevenson also went on to note that, 'Mr Hicks is equally well known as a man of science. He is an authority in his particular subject (geology); and is well versed in other sciences.'

That Hicks was an authority on geology, and 'a force to be reckoned with in the geological community in Britain' by 1877, is not in question, but the question does arise as to the true identity of his geological mentor.

In 1862, the year in which Hicks returned to Wales to serve as doctor and chemist in his native village, John William Salter, the well-known palaeontologist, then employed by the Geological Survey, also paid a visit to St David's. In his note in *Y Bywgraffiadur Cymreig* (The Biographical Dictionary of Wales), F.J. North implies that it was Salter who kindled Hicks' interest in the rocks and fossils of the St David's peninsula. However, unlike North, David Oldroyd was in no doubt whatsoever as to the true identity of Hicks' geological mentor, for in his book *The Highlands Controversy* (1990), a study of the history of geological research in the north-west highlands of Scotland, Oldroyd asserts that 'it was Salter who really introduced Hicks to geology and showed him how to do palaeontological work in the ancient rocks'.

Although Hicks learned much about fossils in the company of Salter, it is doubtful, to say the least, that even somebody of the calibre of Hicks could have communicated such an accomplished paper – his first geological article – documenting his detailed research 'On the Lower Lingula Flags of St David's' before a meeting of the Liverpool Geological Society in December 1863, had he have not been introduced to the subject of geology long before the summer of 1862.

Henry would not have known his great-grandfather Abel Hicks, but prior to the death of his father in 1843 and grandfather in 1845, is it possible, one wonders, that Henry would have had his interest in rocks and fossils kindled to some extent by tales of his great-grandfather who, in 1769, not only established the Folkestone Colliery Company, which worked Folkestone Colliery (SM 8620) near the shores of St Bride's Bay, but also ran a coasting business which shipped coal and culm from the Pembrokeshire coalfield around the

The Bronze Age stone that stands at Tre-maen-hir

coast of Wales and over to Ireland. Furthermore, at Tre-maen-hir he would have been confronted by a Bronze Age standing stone (SM 827263), whilst between Solfach and St David's he would doubtless have become familiar with all the Neolithic burial chambers and early Christian monuments, for which the area is famous. His fascination with the early relics of man cannot be doubted, for less than three years after delivering his first geological paper before members of the Liverpool Geological Society, he was excavating Coygan Cave (SN 284092), between Laugharne (SN 3010) and Pendine (SN 2307), which prompted the publication in 1867 of his first article of geological and archaeological interest.

The identity of Hicks' archaeological mentor is not known – if indeed he ever had one – but it is possible to reveal the identity of the person who really introduced Hicks to geology. It was certainly not Salter, as claimed by so many, but rather the man who placed Salter in touch with Hicks. In a paper written some six years before his death, Hicks recalls that it was 'Through the instrumentality of my relative,

John W. Salter

the Rev. Professor H[enry] Griffiths, of Liverpool (like myself a native of St David's), [that] I was placed in communication with Mr Salter, and in a short time was able, not only to furnish him with better specimens of the *Paradoxides* (giant trilobite) [from Porth-y-Rhaw (SM 786243)] that he had previously obtained, but also to send him other fossils belonging to several new genera'.

Henry Griffiths (1812-91), born at Llanferan a short distance north of St David's, was the son of the Rev. James Griffiths (1782-1858), who was, between 1841 and 1858, the minister of Rhodiad Chapel (SM 767272), near St David's, and Ebeneser Chapel, St David's. James Griffiths was the brother of William Griffiths, of Carmarthen, father of Anne Hicks, Henry's mother. So, Anne Hicks and Henry Griffiths were cousins. Henry, like his father James Griffiths, also entered the ministry and in 1841 was appointed Professor and Principal of the Welsh Independents' Theological College at Brecon. In 1854 he moved to Liverpool, where he spent the next ten years of his life, serving as Dean of Queen's College and minister of Newington chapel.

According to a report that appeared in the *Liverpool Mercury*, 25th December 1854, the unduly modest Rev. H. Griffiths 'did not profess to be a geologist, in the proper sense of the term; nor did he think geology at all his *forte*, though, it certainly had been a favourite study of his for some years. At the very best (in his opinion) he was but a merest amateur, perhaps moderately well versed in the outline, but lamentably deficient in details'. His keen interest in geology is attested by the fact that not only did he hold many evening classes on the subject during his ten-year stay in Liverpool (and, indeed, after leaving the city), but he also led geological excursions. One such field trip, to Hilbre Island (SJ 187877) in the Dee estuary, described in detail in the *Liverpool Mercury,* 26th June, 1855, examined the submerged forest along the Leasowe shore (SJ 2691). A description of one of his other trips to Mynydd Troed (SO 165292) Breconshire was published in 1866. Records of his lectures, including twelve on the

Rev. Professor Henry Griffiths

theme of '*Genesis and Geology*' delivered in the museum of the Liverpool Institute, are recorded in the *Liverpool Mercury* of 25th February, 1858. Not surprisingly, the gifted and talented teacher, became a member of the Liverpool Geological Society and served two terms as its President between 1862 and 1864.

And after having encouraged and inspired Henry Hicks, his cousin's son, to study the rocks and fossils of his native St David's, it comes as no surprise that the Rev. Henry Griffiths chose to launch the geological career of his young protégé by inviting him to address the Liverpool Geological Society during his presidency.

ABSTRACT OF THE PROCEEDINGS

OF THE

LIVERPOOL GEOLOGICAL SOCIETY.

SESSION FIRST.

JANUARY 10TH, 1860.

The PRESIDENT, HENRY DUCKWORTH, F.G.S., F.R.G.S.,
in the Chair.

JOHN PHILLIPS, M.A., LL.D., F.R.S., F.G.S., Oxford; A. C. RAMSAY, F.R.S., London; J. B. JUKES, M.A., F.R.S., Dublin; J. MORRIS, F.G.S., London; S. J. MACKIE, F.G.S., F.S.A., London, were elected Honorary Members.

H. GRIFFITHS, Rev. Prof.; D. CAMERON, M.A.,; CUTHBERT COLLINGWOOD, M.A., M.B., F.L.S.; OWEN LEWIS; WILLIAM G. HELSBY, were elected Ordinary Members.

DECEMBER, 1863.

THE PRESIDENT, THE REV. PROFESSOR GRIFFITHS,
in the Chair.

WILLIAM DAWBARN and E. H. BIRKENHEAD, F.G.S., were elected Ordinary Members.

The following communications were read:

ON THE LOWER LINGULA FLAGS OF ST. DAVID'S, PEMBROKESHIRE.

BY HENRY HICKS, M.R.C.S.E.

HAVING lately examined the fossiliferous beds of the Lower Lingula Flags in the neighbourhood of St. David's, Pembrokeshire, and with, I am happy to say, a certain amount of success, I thought it would not be unacceptable to the members of this Society to forward a few specimens of the fossils found there, for presentation to the Museum, and to accompany the same with a short analysis of the beds, as well as to give a few facts concerning their relative position, &c.

Indeed, there can be little doubt that, thanks to Henry Griffiths, Hicks was a keen amateur geologist before his student days in London. Furthermore, in London he was able to increase his knowledge of the subject by attending lectures held at the headquarters of the Geological Society of London, in Somerset House, less than two miles from Guy's Hospital. And what an exciting time it was, for in 1859, the year that Hicks embarked on his medical studies, Charles Darwin, himself a former secretary of the Geological Society, published *The Origin of Species*. Furthermore, it was in Snowdonia that Darwin undertook his geological apprenticeship in the 1830s and 40s.

By the time Henry Griffiths introduced Salter to Hicks in 1862, there is little doubt that the Surgeon at St David's was well-acquainted with the ancient strata so magnificently exposed between Solfach and St David's Head (SM 720278). Indeed, in a paper communicated to the Geological Society of London in 1864, the year in which Hicks was elected an Honorary Member of the Liverpool Geological Society, Salter generously acknowledges the help of both Hicks and Henry Griffiths: '. . . by the co-operation of Professor Griffiths, of Liverpool, and particularly the zealous examination of the beds by Mr. Henry Hicks, Surgeon at St David's, we are now in possession of much more abundant and perfect materials'. And Hicks was also deeply indebted to Salter, as he made clear in his communication to the Liverpool Geological Society in December 1863: '. . . I must publicly acknowledge my great obligations to Mr. Salter, who has kindly, from time to time, named and classified the fossils as they were found by me, and to whom I am indebted for much other information concerning these beds [i.e. the rocks of Porth-y-Rhaw]'. In fact, one of Salter's major discoveries was an enormous, nearly two feet long trilobite, *Paradoxides davidis,* at Porth-y-Rhaw, which he named after his friend, David Homfray.

To seal their friendship, Hicks and Salter, both named fossils after one another: Hicks, in 1865, naming the trilobite *Anopolenus*

Porth-y-Rhaw

Salteri after 'My friend Mr. Salter . . . in memory of the pleasant days spent together on the cliffs of St David's', and Salter, returning the compliment by naming the trilobite *Paradoxides Hicksii* in honour of 'my friend Mr. Hicks'.

Not only did Hicks and Salter enjoy 'pleasant days spent together on the cliffs of St David's', which resulted in the production of some very fruitful research, but also at the 1867 National Eisteddfod of Wales held in Carmarthen, Hicks' mother's home town. The Geological Exhibition, part of the Science Exhibition of the Eisteddfod held in the Carmarthen Assembly Rooms, provided Hicks, in particular, with an opportunity to exhibit some of his exciting finds. The local newspaper, *The Welshman*, tells us that '. . . Mr Hicks, of St David's, shows specimens from the Cambrian rocks of that district, which are not only entirely new and unique in themselves, but of the utmost importance to science, defining a new *fauna*, older than even our highest authorities have been prepared to accept. ... The bone caves near Tenby and Laugharne [Coygan] are represented by the very excellent collections

*Trilobite: Conocoryphe applanata,
from the Hicks Collection, Department of
Geology, University Museum, Oxford*

of the Rev. G. Smith, of Gumfreston (SN 109011), and Mr. Hicks, of
St David's . . .' And one suspects that it was Hicks who arranged for
Salter to deliver a lecture at the same Eisteddfod extolling the virtues of
studying the sciences in general and geology in particular.

But the Hicks – Salter relationship was also one of doctor and
patient. Unable to work harmoniously with his fellow officers in the
Geological Survey, Salter was forced to leave his post in July 1863 and
from that time onward his 'difficult temperament was evidently giving
way to insanity. He was frequently a very ill man in both body and
mind' and Hicks, who had a particular interest in mental diseases,
took it upon himself to care for 'poor Salter', as he invariably referred
to him in letters addressed to other geological friends. Indeed, given
Salter's plight, there can be little doubt that the Reverend Professor
Henry Griffiths had more than geological companionship in mind
when he introduced the ailing palaeontologist to Dr Hicks. Sadly,
Hicks failed to quell Salter's troubled mind and two years later, in
August 1869, Salter committed suicide.

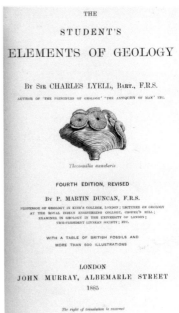

THE

STUDENT'S

ELEMENTS OF GEOLOGY

By Sir CHARLES LYELL, Bart., F.R.S.

AUTHOR OF 'THE PRINCIPLES OF GEOLOGY' 'THE ANTIQUITY OF MAN' ETC.

Thecosmilia annularis

FOURTH EDITION, REVISED

By P. MARTIN DUNCAN, F.R.S.

PROFESSOR OF GEOLOGY IN KING'S COLLEGE, LONDON ; LECTURER ON GEOLOGY
AT THE ROYAL INDIAN ENGINEERING COLLEGE, COOPER'S HILL ;
EXAMINER IN GEOLOGY IN THE UNIVERSITY OF LONDON ;
VICE-PRESIDENT LINNEAN SOCIETY ; ETC.

WITH A TABLE OF BRITISH FOSSILS AND
MORE THAN 600 ILLUSTRATIONS

LONDON
JOHN MURRAY, ALBEMARLE STREET
1885

The right of translation is reserved

Sir Charles Lyell

Despite the loss of his friend, Hicks' research work amongst the ancient rocks of his home area continued unabated. His confidence, and ego, had also been given an enormous boost in September 1868, when Sir Charles and Lady Lyell visited Henry Lyell, Charles' brother, who lived at Tenby (SN 1200). During their stay, Charles Lyell (1797-1875), author of the highly influential book, *The Principles of Geology*, published in three volumes between 1830 and 1833, took his nephew Leonard to St David's in order to 'examine the magnificent sections of fossiliferous Cambrian rocks, under the guidance of Hicks'. After the visit, Lyell, in a letter to W.S. Symonds, recalls enthusiastically the 'very successful expedition to St David's, where Dr. Hicks devoted the best part of three days to show us some 2,000 feet of strata'.

As for Hicks, he was over the Moon. In a letter dated 1st October 1868 he thanked Sir Charles for the gift which he had received from the great man: 'I feel exceedingly obliged to you for the two beautiful vols [unfortunately, their titles are not disclosed] which you have kindly sent me. – I shall indeed value them much, as having been presented by the author . . . Your 'Elements' [i.e. *Elements of Geology*, first published in 1838, the year after Hicks was born] was the *first geolog.* work ever read by me, therefore it is not to be wondered at, that I should have felt so honoured at having the pleasure of taking you over our *geolog.* ground and showing you the fruits of our comparatively insignificant labours.'

From that time onwards, and up to within a year of Lyell's death in 1875, Hicks was in regular correspondence with Lyell, the author of *Principles of Geology*, the celebrated volume which greatly influenced Darwin's ideas on evolution as he sailed around the world on the *Beagle*. Furthermore, Lyell was greatly impressed by Hicks, for he often wrote to him requesting information relating to the rocks of the St David's area and seeking his opinion on other geological questions. Not surprisingly, therefore, Hicks' name frequently appears in the text and index of Lyell's *Student's Elements of Geology*, first published in 1871, the year in which Henry Hicks and his wife and two daughters left Glasfryn (SM 758254), the house in St David's that he had had built for himself and his family, to settle in Hendon, a village (at the time) situated some 7 miles north-west of London.

Prior to 1871 Henry Hicks had only spent three years away from home. Following his college days, his home for two years was with his mother and sister in the house adjacent to the Druggist's House on Cross Square, St David's (SM 753253) which his sister looked after on those occasions when Hicks was visiting patients or searching for fossils.

In the early days, this was also the location of Hicks' surgery. On 2nd February, 1864 Hicks married Mary, the only daughter of the Rev. P.D. Richardson, vicar of Llantydewi (SM 968280), Pembrokeshire and the couple moved to Glasfryn, St David's where two of his three daughters, Annie Elizabeth (1866) and Minnie Jane (1870), were born.

Hicks' reasons for leaving Glasfryn and St David's are not altogether clear but Professor Thomas George Bonney (1833-1923),

The Druggists' House, St David's

a friend of Hicks and author of his obituary notice published in the *Proceedings of the Royal Society of London*, offers the following thought-provoking comment: 'At that date [i.e. the 1860s, the decade following the publication of Darwin's *The Origin of Species*] geology was not favourably regarded in clerical and county circles and to be suspected of it might have been injurious to the young practitioner. So, as Hicks used to relate in later years, he had to carry on geological investigations behind the screen of professional duties, and his servant must have often wondered at his master's fondness for leaving his carriage to walk home across the moor or along the cliffs.'

SAINT DAVID'S.

MR. DAVID PERKINS

Has been favoured with instructions from H. Hicks, Esq.,
(who is leaving the neighbourhood), to

SELL BY AUCTION,

AT HIS RESIDENCE,

GLASFRYN HOUSE,

SAINT DAVID'S,

On Thursday, March the 9th, 1871,

Although interesting, it's hard to believe that this was Hicks' only reason for abandoning his home, general practice and chemists' shop. After all, following his student days, he was more than aware of the fact that it would be far easier for him to further his burgeoning geological career in London, than in the remote and tiny city of St David's. Early in 1871, Hicks and his young family settled in Heriot House, Hendon (TQ 233888), at that time a village to the north-west of London. At Hendon, he would be less than ten miles away from the headquarters of the three most important geological institutions in Britain: the Geological Survey, based at the Museum of Practical Geology, Jermyn Street, Piccadilly; the Geological Society, based at Somerset House, and Burlington House, Piccadilly, from 1874 onwards; and the Geologists' Association, which usually held its meetings at University College, Gower Street.

Once settled in his new home, Hicks lost no time in furthering his geological career: by November 1871, Dr Henry Hicks of Heriot House, Hendon, had been elected a Fellow of the Geological Society of London after having been recommended 'as a proper person' by two signatories, Sir Charles Lyell and W.S. Symonds, author of *Records of the Rocks* (1872), a volume containing several references to Hicks' research work in the sections referring to the rocks of the Pre-Cambrian and the Cambrian periods. At about the same time he also became a member of the Geologists' Association.

From his home in Heriot House, Hendon, Hicks worked initially as a general practitioner and in February 1879 established 'Hendon Grove Asylum', where he, according to the 1881 census, together with his wife, daughters and staff cared for 12 mentally ill female patients. In addition to caring for his patients, Hicks was also very active in local affairs, especially in questions sanitary and educational, as well as in church work, and the affairs of the Conservative party. Hicks was also a faithful, conscientious and well-known member of the London Geological Society and the Geological Association, where he enjoyed further discussions and debates. Meetings of the Hendon

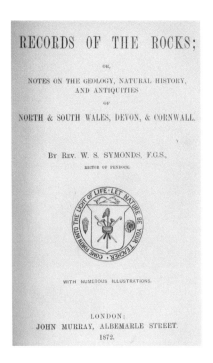

RECORDS OF THE ROCKS;

OR,

NOTES ON THE GEOLOGY, NATURAL HISTORY,
AND ANTIQUITIES

OF

NORTH & SOUTH WALES, DEVON, & CORNWALL.

BY REV. W. S. SYMONDS, F.G.S.,

RECTOR OF PENDOCK.

WITH NUMEROUS ILLUSTRATIONS.

LONDON:
JOHN MURRAY, ALBEMARLE STREET.
1872.

Debating Society, where, true to character, 'he greatly enjoyed a sharp skirmish of words', provided him with a break from medical and public duties. As did his geological investigations, for although Hicks had settled in Hendon he had by no means abandoned his work in Wales, nor did he miss an opportunity to further his geological knowledge by visiting and undertaking valuable research in locations far beyond the Welsh border.

Consider, for example, his schedule during the 1870s. Hicks spent time examining the sedimentary and igneous rocks of Ramsey Island (SM 7023) in 1872 and again in 1873. In August 1874 Hicks found in the quarry at Llan-fyrn [Llanvirn *sic*] (SM 797302), near Abereiddi (SM 798316), interesting and important fossils. August 1875 was spent searching for fossils in the Pen-y-Glog slate quarry, near Corwen (SJ 0743), although he also allocated time in 1875 and 1876 to study in greater detail rocks in and around St David's. During the summer of 1877 he spent a further three weeks researching the rocks of St David's and a period of time in Snowdonia. In 1878, he returned to Snowdonia, but also undertook research on Anglesey, in the Loch Maree (NG 9172) area of Ross-shire in the north-west highlands of Scotland, St David's and along the coast of Dublin and Wicklow! In the closing year of the decade he familiarised himself with the geology of the Malvern Hills (SO 7646) and the Rhinog Mountains of north-west Wales (SH 6527).

Prior to 1870 much of Hicks' time had been devoted to the search for fossils. Indeed, he seemed almost to scent out fossils from even the most unpromising strata, and by as early as 1868 had discovered thirty species, mostly trilobites, in the Cambrian [rocks] of St David's. After 1870, however, he focussed his attention on establishing whether Archaean (i.e. Precambrian) rocks older than the fossiliferous strata of Cambrian age were to be found not only in the St David's area, but also elsewhere in Wales, work which brought him into direct conflict with the geological establishment, resulting in many a 'sharp skirmish of words'.

Indeed, Hicks became embroiled in an acrimonious, tripartite conflict, which became one of the most severe disputes in British geology. The establishment of the Geological Survey of Great Britain, effectively in the year 1835, had for the first time brought into being a band of professional geologists, who, like so many other so-called 'professionals', were reluctant to acknowledge that the work of 'amateurs', such as Henry Hicks, was worth the paper on which it was written. Between the two opposing poles, 'professional' and 'amateur', was a new breed of 'academic' geologists, the product of the universities, men such as Thomas George Bonney, a graduate of Cambridge University and Professor of Geology in University College, London who, incidentally, in his presidential address delivered before the Geological Society in 1885 ventured to express the wise counsel that 'a man who has devoted his life to the study of geology is equally a professional geologist'.

University College London (1876)

By advocating the presence in Pembrokeshire of rocks of
Archaean/Precambrian age, a concept which the professional
geologists of the Geological Survey of Great Britain did not entertain
in Britain at the time, it was inevitable that 'a sharp skirmish of words'
loomed on Hicks' horizon. In fact, he was destined to incur the
wrath of Sir Andrew Ramsay who, in 1871, succeeded Sir Roderick
Murchison as Director General of the Geological Survey, and, more
particularly, Sir Archibald Geikie, who succeeded Ramsay in 1881.

As early as 1864 Salter and Hicks had suspected the presence
at St David's of a ridge of Pre-Cambrian rocks. Hicks drew attention
to this in 1872, and referred more particularly to the subject in a
paper read before the Geological Society, in its new headquarters in
Burlington House, in December 1874. Condescendingly, Director
General Andrew Crombie Ramsay, who was appointed in 1871
following the retirement of Sir Roderick Murchison, complimented
Hicks on his paper but, after referring to the findings of his own
work in Pembrokeshire, which he commenced in 1841, he scathingly
enquired of the 'amateur' upstart, 'How old were you at that time?'

Sir Andrew Crombie Ramsay

and added sarcastically, 'Why, you were only a baby!'

Hicks, always 'warm-tempered . . . [and] ever ready to do battle', was 'undeterred, even by a voice as authoritative as that of the Director General of the Survey'. The truth is, his 'heretical' ideas were gaining a degree of acceptance elsewhere in Britain. However, a hint that the wranglings over the interpretation of the rock record in the St David's area, which had rumbled on throughout the 1870s, were about to come to a climactic head is to be found in a letter written by Edward Bernard Tawney (1841-82) of the Woodwardian Museum, Cambridge, to Archibald Geikie, at about the time of his ascension to the Director General's throne.

With reference to some of the rocks of the Llŷn peninsula, an area familiar to Hicks, an irritated Tawney complains: '. . . I have so far given in to Hicks as to suppose provisionally that [the rock] is metamorphic, but that is not admitting any Pre-Cambrian age . . . it is about time that the Hicks bubble should burst,' exclaimed the exasperated Tawney, adding that Hicks' hypothesis regarding the age of the rocks of Llŷn '. . . is simply a preconceived theory bolstered up by merely one carriage drive through the country and confirmed by a complete ignorance of the elements of petrology and mineralogy'!

Sir Archibald Geikie

Sir Archibald Geikie, disliked even amongst several of his Survey colleagues, for his arrogance, had been primed. He was determined to burst 'the Hicks bubble'. But in truth, he hardly needed prompting, for Hicks, following his visit to the Loch Maree area of north-west Scotland in 1878, published a paper criticising the findings of the authors of an earlier paper entitled 'On the altered rocks of the western islands of Scotland, and the north-western and central Highlands' (1861). The authors in question were none other than Roderick Impey Murchison, Director of the British Geological Survey, and Archibald Geikie, Local Director of the Geological Survey in Scotland. Hicks' excursion into Geikie's Scottish territory was, almost certainly, a deliberate act of provocation. For Hicks, it was an opportunity to challenge the 'professionals' and the 'geological establishment', and he surely knew that his iconoclastic paper, which effectively kindled the 'Highlands Controversy', would do more than ruffle the feathers of Geikie. Indeed, following Hicks' presentation, Ramsay did his utmost to defend the standpoint of the Geological Survey but his attempts were unsuccessful. In one fell swoop Hicks had hit the three influential Scotsmen with one stone!

Sir Roderick Impey Murchison

Hicks was at it again in February 1883 when he delivered, before the assembled members of the Geological Society, another paper on the rocks of Ross-shire and Inverness-shire, based upon further research in the area in 1880 and 1881.

By trespassing unreservedly on Geikie's research area and by putting pen to paper, Hicks made a positive contribution to the the geological world, in addition to the scientific world in general, by showing that the opinions and theories of an amateur geologist could draw attention to weaknesses in the official explanations and successfully challenge the assumptions and prejudices of some professional geologists.

In order to undermine Hicks' work Geikie paid his first visit to St David's, in the company of Benjamin Peach, in September 1882, (Peach and Horne were instrumental in the mapping and deciphering of the Moine Thrust, Scotland), and then again in February 1883. On the basis of these two visits, Geikie presented a vitriolic attack on Hicks' fieldwork and findings during the meetings of the Geological Society in March and April 1883. Unsurprisingly, Geikie's paper was entitled 'On the supposed Precambrian rocks of St David's'. In presenting his research, Geikie took the opportunity to attack viciously the work of Hicks. The geological details of the arguments, and the heated discussions, which took place in the debating chamber of the Geological Society between the rival factions – the 'professionals' spearheaded by Geikie, and the 'amateurs/academics' led by Hicks – have been admirably recorded by David Oldroyd in his paper 'The Archaean Controversy in Britain: Part I – The Rocks of St David's', published in *Annals of Science* in 1991 and by Paul Pearson and Christopher Nicholas in their paper 'Defining the base of the Cambrian: the Hicks - Geikie confrontation of April 1883', published in *Earth Sciences History* in 1992. Suffice it to say that Geikie was in no mood to give up the fight. He was determined to quosh the implied incompetence of Geological Survey staff and defend the Survey's orthodoxy, namely that there were no Precambrian rocks in the St David's area. Indeed, as a parting shot at the close of the meeting, Geikie patronisingly expressed the hope that 'Dr Hicks and himself would continue friends, and that a time would come when they would go over the ground together neither believing in the existence of Precambrian rocks'.

Hicks, unsurprisingly, was not prepared to throw in the towel. His response was to seek the aid of fellow Archaeans – [Thomas McKenny Hughes, Professor of Geology at Cambridge University, and Thomas Bonney, Professor of Geology at University College, London] – in the field in the hope that they would help Hicks prepare his counter-attack, which, ironically, he launched in 1884, the very year in which the Geological Survey, under Geikie's leadership, had to admit to the errors of its interpretation of the rock succession in the north-west highlands of Scotland. However, despite the fact that Peach and Horne, of the Geological Survey, had demonstrated to Geikie in the field that the work of Charles Calloway and Charles Lapworth, two 'amateurs' like Hicks, was correct, Geikie steadfastly declined to acknowledge their work. Hicks, on the other hand, in his counter-attack of 1884 did yield on some matters raised by Geikie relating to his interpretation of the rocks of the St David's peninsula but he did not budge on the central point at issue, namely that rocks of Precambrian age were to be found there. In the fullness of time, Hicks was proved correct and, before he died, he could take some satisfaction from the fact that even Geikie had to concede the principle that he had flatly denied for so many years that 'there could be Precambrian rocks in England and Wales, even if St David's was still excluded'.

Given Geikie's abortive attempt 'to burst the Hicks bubble', it comes as no surprise to learn that neither his visits to St David's in 1882 and 1883, nor his verbal battles with Hicks are recollected in his autobiography, *A Long Life's Work*, published in 1924. Neither was Hicks' vital contribution to the Highlands Controversy, which duly inspired the work of Ben Peach and John Horne and resulted in the publication of the *The Geological Structure of the North-west Highlands of Scotland*, 1907, acknowledged. However, although the relationship between Geikie and Hicks was acrimonious, the correspondence between the two men was unavoidably polite and courteous after Hicks was appointed Secretary of the Geological Association in 1890.

Ironically, however, another battle between the two combatants seemed to be looming on the horizon, following an argument which had arisen between McKenny Hughes and Hicks, both fellow Welshmen and fellow Archaeans. In 1883, when the Hicks – Geikie confrontation was at its height, Hicks paid his first visit to Ogof Ffynnon Beuno and Ogof Cae Gwyn (SJ 015710), near Tremeirchion on the eastern flanks of the Vale of Clwyd. Realising the potential importance of the site as regards the Pleistocene history of the area and questions relating to the antiquity of man, Hicks embarked upon a series of excavations during the summers of 1884, '85, '86 and '87. He was also instrumental in organising a small excavation committee, of which he was secretary and of which McKenny Hughes was a member.

Ogof Ffynnon Beuno and Ogof Cae Gwyn

Hicks published about ten articles reporting on the findings of the excavations and interpreting the significance of the finds. Indeed, Hicks published convincing evidence which demonstrated: firstly, that 'the bone bed', traceable beyond the mouth of the cave, included not only the bones of hyaena, rhinoceros, horse and mammoth but also flint implements; and secondly, that it was overlain by glacial deposits of Irish Sea derivation. But it's clear that the question concerning the antiquity of man troubled McKenny Hughes, the son of the Bishop of St Asaph, and he could not accept that prehistoric man had wandered the Vale of Clwyd prior to the Glacial Period, as claimed by Hicks. Perversely, he argued that the flint implements and the mammalian fauna were of Post-glacial age.

But the argument concerning the antiquity of man was not the only bone of contention between Hicks and Hughes. Hughes, as far as Hicks was concerned, proved to be an unreliable ally in his fight against Geikie; in fact, in the discussion that followed Hicks' 1884 counter-attack he questioned the validity of some of the evidence presented by Hicks. And it now seemed as though Hughes himself was determined to 'burst the Hicks bubble'. Clearly, the services of an arbiter were called for to settle the argument between the two erstwhile friends. By now, Hicks was brimming with confidence and in October 1887, at the request of both Hughes and himself, he agreed that Archibald Geikie, of all people, should settle the dispute. If Hughes had hoped 'to burst the Hicks bubble' with the aid of Geikie, he too was destined to be disappointed for in his report the Director of the Geological Survey found himself obliged to dissent from the Professor's views and, instead support Hicks' interpretation; an interpretation which has subsequently assumed far greater importance following the carbon-dating of one of the mammoth bones.

By now, Hicks was riding the crest of a wave. In September 1888, at the International Congress of Geologists, held at the University of London, it was Hicks who was invited to open the discussion on Cambrian-Silurian classification. For this prestigious event he produced a document which Mike Bassett and Ellis Yochelson rightly described as a 'prescient map' of the geology of north Wales which distinguishes for the very first time, in time and space, the three divisions – Cambrian, Ordovician and Silurian – of the Lower Palaeozoic. The 'Ordovician system', fully endorsed by Hicks, had been proposed by his good friend and fellow 'amateur' Charles Lapworth in 1872 but the term did not gain the seal of approval of the 'professional' geologists of the Geological Survey until the first decade of the twentieth century.

Hicks was also responsible for planning, directing and leading the post-conference excursion across north Wales, during which time he had the honour of the company of the great American palaeontologist Charles Doolittle Walcott (1850-1927). Walcott was the palaeontologist and stratigrapher famous for his discovery of the unique fossils within the 500 million year old rocks of the Burgess Shale, high in the Canadian Rockies. In fact, the American and Welshman were birds of a feather for Walcott, like Hicks, had had no formal training in geology.

The 1880s had proved to be a significant decade in the life of Henry Hicks. In addition to the events already alluded to, Hicks, in 1883, in the year he was elected President of the Geologists' Association, a position which he held until 1885, was also presented with the solid gold Bigsby Medal, awarded by the Geological Society, 'for work of great merit' and 'as an acknowledgement of eminent services in any department of Geology'.

And then, in 1885, came the ultimate accolade. 'I desire on behalf of myself and my colleagues,' said a Mr Hancock, member of the Hendon Drainage Committee, in a meeting held in early May 1885, 'to offer our hearty congratulations to our friend Dr. Hicks, the chairman . . . on the distinction conferred upon him since our last meeting in having been elected a 'Fellow of the Royal Society' – a distinction which there is no doubt he has well earned.' In response, the immodest Hicks stated 'that it had been his aim to obtain the position, and that he had received the distinction at an unusually early age.'

The Bigsby Medal

Doubtless the forty-eight year old doctor would not have reacted kindly had he have been reminded that Archibald Geikie was elected Fellow of the Royal Society at the tender age of thirty! And finally, in 1896, Hicks was elected president of the Geological

Society, after having served as its secretary between 1890 and 1893, and member of the Council on a number of occasions between 1875 and 1899, the year he died. His election to the presidency of both the Geologists' Association and the Geological Society of London is an indication that his contribution to geological research in general – in south-west Wales (Pembrokeshire and Carmarthenshire), north-west Wales, north-east Wales, the north-west highlands of Scotland, north Devon, and in the vicinity of London – was much appreciated and recognised by his peers.

That Hicks was a deserving recipient of the honours that came his way, there can be no doubt, but *heb ei fai, heb ei eni*: he who is faultless, has not been born. Hicks was, without doubt, immodest, arrogant, quick-tempered, and 'In drawing conclusions he was very quick – too quick . . . in publishing them'. But he was also a bundle of 'unflagging energy, and seemingly abundant vitality', a man possessed of enthusiasm who found in geology a 'means of recreation and of much intellectual enjoyment'. Judged by the standards of his day, his researches were a significant contribution to the understanding of the geology of those areas on which he had focussed attention. Had he have not died of heart disease at the age of 62, we are left wondering whether Hicks, who introduced into the geological literature for the first time names such as Pebidian, Dimetian and Llan-fyrn, still part of the geologists' vocabulary, would also have published a book entitled *The Oldest Rocks in the British Isles*. A copious set of notes deposited in the National Library of Wales, Aberystwyth, in 1911, by his daughters, Mrs Annie Elizabeth Knethell Green, Glan-y-môr, St David's, and Mrs Ethel Mary Richardson, Cilau-wen, Letterston, certainly suggests that this was his intention. Much of it would have been devoted to describing the Precambrian rocks of Wales. Although Hicks was a Welsh speaker his notes are written in English, a symptom of the Victorian age when even the annual National Eisteddfod of Wales promoted the use of English particularly for technical and scientific subjects such as Geology.

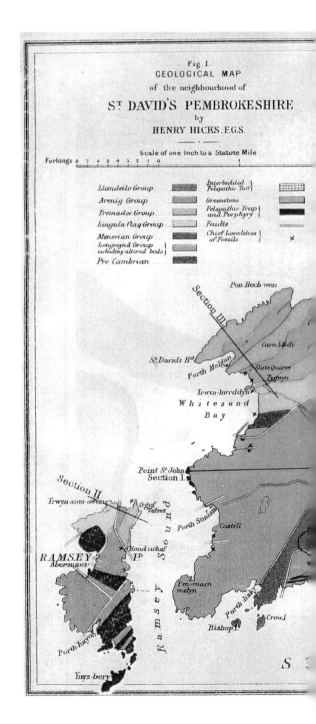

Fig. I.
GEOLOGICAL MAP
of the neighbourhood of
St DAVID'S PEMBROKESHIRE
by
HENRY HICKS. F.G.S.

Scale of one Inch to a Statute Mile

Furlongs 8 7 6 5 4 3 2 1 0

Llandeilo Group.
Arenig Group
Tremadoc Group
Lingula flag Group
Menevian Group
Longmynd Group
including altered beds
Pre Cambrian

Interbedded
Felspathic Tuff
Greenstone
Felspathic Trap
and Porphyry
Faults
Chief Localities
of Fossils

Pen-llech-wen
Section III
Carn Lkidi
St Davids Hd
Porth Melsan
Slate Quarry
Tygwyn
Trwyn-hwrddyn
Whitesand
Bay

Point St John
Section I.

Section II
Trwyn-sion-owen
Oggt
Velvet
Porth Stinian
Castell
Road uchaf
Pd
RAMSEY
Abermawr
Pen-maen
melyn
Porth lisky
Porth Lysog
Bishop I.
Crow I.
Ynys-bery

Ramsey Sound

S

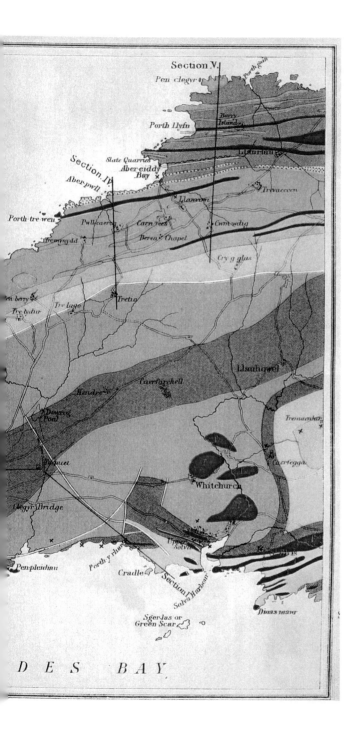

Section V.

Pen clegyr

Porth gain

Barry Island

Porth Llyfn

Llanrian

Slate Quarries

Section IV.

Aber-eiddy Bay

Aber-pwll

Trevaccoon

Llanwn

Porth-tre-weu

Pwllcaerog

Carn rees

Cwm-wdig

Tremynydd

Berea Chapel

Cry g glas

n berry

Tre iago

Tretio

Tre tydur

Llanhowel

Caerfarchell

Hendre

Tremaenhir

Dowrog Pool

Caerfegga

Trefinel

Whitchurch

Clegyr Bridge

Upper Solva

Penpleidian

Porth-y-rhaw

Cradle

Section Solva Harbour

Dinas mawr

Sger lus or Green Scar

D E S   B A Y

St David's Church, Tre-groes/Whitchurch, Pembrokeshire

Henry Hicks died on 18 November 1899 and was buried, not amongst members of his family, his father, mother and brother – Thomas, Anne and John Hicks – and other members of the Hicks family of St David's, at Tre-groes/Whitchurch, near Solfach, but in Hendon churchyard, adjacent to St Mary's parish church. Sadly, his gravestone no longer exists. But, far from his native St David's, Hicks is commemorated, for affixed to a wall inside St Mary's Church, Hendon, is a small, inconspicuous, rectangular tablet of alabaster, measuring 46cm by 30cm, which reads 'In ever loving memory of Henry Hicks Esq MD:FRS of Hendon Grove in this Parish . late President of the Geological Society of London who died Novr 18th 1899 aged 62 years'.

Other than obituaries that appeared in scientific journals an entry in *Y Bywgraffiadur Cymreig* (1953), written by F.J. North, a former keeper of the Geology Department of the National Museum of Wales, Aberystwyth is the only public record of the life and work of Henry Hicks. As author, and sometimes co-author, of over 60 articles and 30 reports, short papers and letters relating to Geology and Geomorphology this pioneering geologist deserves greater attention. If Hicks had been a poet or novelist, a musician or minister, somebody, long ago, would have ensured a suitable memorial to him in St David's.

Hicks, one of St David's most influential sons, doctor and chemist, accomplished 'amateur' geologist and Welsh-speaking Welshman deserves, at the very least, a permanent memorial in St David's, nothing less than a plaque to be placed on the walls of his old chemists' shop in Cross Square, bearing the following bilingual inscription: *Yma, rhwng 1862 a 1870, y gweithiai Dr Henry Hicks, MD, FRS (1837-99), brodor o Dyddewi, meddyg a fferyllydd, a daearegydd o fri. Here, between 1862 and 1870, worked Dr Henry Hicks, MD, FRS (1837-99), a native of St David's, doctor and chemist, and geologist of distinction.*